Science Sensations

Science Sensations

An Activity Book from
The Children's Museum, Boston

Diane Willow and Emily Curran

Illustrated by Lady McCrady

Addison-Wesley Publishing Company

Reading, Massachusetts • Menlo Park, California • New York
Don Mills, Ontario • Wokingham, England • Amsterdam • Bonn
Sydney • Singapore • Tokyo • Madrid • San Juan
Paris • Seoul • Milan • Mexico City • Taipei

Library of Congress Cataloging-in-Publication Data

Willow, Diane.
    Science sensations / Diane Willow and Emily Curran;
illustrated by Lady McCrady.
        p.   cm. — (A Boston Children's Museum activity book)
    Includes index.
    Summary: Activities/experiments to give a new
awareness of the world around us. Treats such areas as
light, color, shadows, reflections, motion pictures, illusion,
and patterns.
    ISBN 0-201-07189-4
    1. Science — Experiments — Juvenile literature.
[1. Science — Experiments.   2. Experiments.]
I. Curran, Emily.   II. McCrady, Lady, ill.   III. Title.
IV. Series.
Q164.W544   1989
507'.8 — dc19   88-27226   CIP   AC

Cover illustration by Dianne Cassidy
Cover design by Copenhaver Cumpston
Text illustrations by Lady McCrady
Text design by Joyce C. Weston
Set in 11½ point Egyptian by DEKR Corp., Woburn, MA

3 4 5 6 7 8-CRS-97969594
Third printing, May 1994

With love to:
My Mother with her sense of wonder,
Joanne with her courage to question,
and Zora with her joy in being.
— *D. W.*

To my sensational family — Gloria, George, and
Becky Curran — and especially John Callahan.
— *E. C.*

Contents

Acknowledgments

Many of the activities in this book were developed — and all were refined — in the process of our work in the Community Outreach Program at *The Children's Museum, Boston*. The inspiration for many of them comes from the natural world, a continual source of wonder, beauty, variety, and change. The materials we used catalyzed new ideas, as did our colleagues, educators who have come before us, and the children who worked with us. The enthusiasm of hundreds of children who participated in these activities guided us in the choice of what to include in this book. Their involvement enabled us to present activities that really work with children, are fun, and become truly enhanced by each person's particular style of exploration and expression.

We would especially like to thank the staff and children at the following places: in the South End, Youth Center at Villa Victoria and Ellis Memorial Center; in Roxbury, Bridge Fund Inc. Afterschool Program and the Afterschool Program at Alianza Hispana; and in Dorchester, the Raphael Hernandez School.

In addition our appreciation to Pat Steuert, Dottie Merrill, and Suzanne Le Blanc of The Children's Museum, Boston, who helped get the project off the ground. Thanks to our colleagues past and present at The Children's Museum, Boston: Amica, Esther Kohn, Ilya Pratt, Jeff Winokur, and Bernie Zubrowski.

And thanks to our friends for their moral support: Betsy Allen, John Callahan, Gloria Curran, Kim DiMauro, Robert Meek, Joanne Rizzi, Sue Robbins, Jill Wolhandler, and finally a special thanks to each other.

Introduction

Every day amazing things happen. They can be as incredible as a rainbow suddenly appearing after a rain shower, or as simple as a feather floating by. Whether we realize it or not, we're surrounded by science sensations.

This book invites you to experience the world in new and different ways. It's filled with activities that reveal the extraordinary discoveries you can make about ordinary things: how to use shadows to put on a play, or make a rainbow with sunlight; how to decode invisible messages with grape juice, or make your own ink from grass; how to make bubbles last longer, or how to flatten a drop of water.

Whatever chapter you choose to explore — Light, Color, Shadows, Reflections, Water, Wind, Balance, Illusions, Moving Pictures, or Patterns — you'll find activities that are fun and easy to do. You don't need any special equipment. You can do most of the activities by yourself, although some are even more fun when you do them with a friend. You don't even have to know a lot about science! Once you've mastered some of these sensational activities, you can amaze your friends and family with your new knowledge and talents.

Each activity has been kid-tested and kid-approved — and each one really *does* work. Best of all, you can do these activities over and over again because each time you try them you'll learn something new. You may even invent some science sensations of your own.

LIGHT

Light does more than allow us to see. It lets us view the world not only in black and white but in all the hues of the rainbow.

We usually think of light as colorless or white, but in fact light is filled with color. Imagine an ordinary sunny day with a blue sky, red brick buildings, and green trees. As the light fades into dusk everything seems to lose its brilliant color and become muted and gray. That's because we need light in order to be able to see the color of things. Most objects don't produce light. They reflect light that falls on them, and we see the reflected light with our eyes. As mind-bending as it may seem, the color of an object depends on the color of light that it reflects back into our eyes.

Homemade Rainbows

If you're lucky, you may have seen a rainbow arching across the sky after a rainstorm. You may also have seen a rainbow while watching water from a spraying hose, fountain, or rain shower. To see a rainbow you need to have your back to the sun and face the rain or water spray.

No two people ever see exactly the same rainbow. It looks different from different angles. Sometimes if you move it disappears, and then you need to search for that certain angle that makes the rainbow reappear.

Although rainbows are indeed rare in nature, you don't have to be a wizard to conjure one up. Here's how you can make a rainbow appear whenever you are in the mood to be dazzled.

You will need:

a shallow pan, such as a baking dish or a
    brownie pan
water
a pocket mirror
a sheet of white paper or white cardboard
a sunny day

Ghostly Colors

"Spectrum" is a word invented by Sir Isaac Newton (1642–1727), a British scientist, for the band of colors that appears when light is bent. He called it "spectrum" after the word "specter," which means a ghost or ghostly vision.

Place the pan on the floor or on a table, directly in the sunlight. Fill it with water, not quite to the top.

Place the mirror faceup in the water, at an angle, as shown.

Now take your sheet of paper or cardboard and move it around in front of the mirror until you catch a "rainbow" on it. This rainbowlike band of colors is called the *spectrum*.

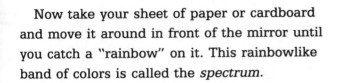

You can change the shape of your spectrum by moving the sheet of paper that it's on. Can you change it by moving the mirror?

Sunlight moves in straight lines, but sometimes, when it passes through something else — water, for instance — the lines of light are bent. When they are bent just the right amount, they fan out into a band of colors called the spectrum. The lines of sunlight that were reflected off your mirror were bent enough as they passed through the water in your pan, and created a spectrum.

There are seven colors in the spectrum, and they always appear in the same order: red, orange, yellow, green, blue, indigo (blue-purple), and violet. Can you find these seven colors in your spectrum? You'll need to look carefully, because each color seems to blend into the next.

A rainbow in the sky is a giant spectrum. It appears when sunlight is bent by the tiny water droplets that linger in the air after a rain shower. The bent light separates into the rainbow colors.

New Worlds to See

What if you looked out the window one day and the grass was black, the sky was purple, and the clouds were red? You can actually see such a scene — and experience the world in a completely new way by filtering light with colored plastic.

You will need:

two sheets of transparent plastic, one red and one green (colored plastic report covers are perfect for this activity; you can buy them wherever stationery is sold)

two crayons or markers, one red and one green

a sheet of white paper

Hold the red plastic in front of your eyes and look through it. What color is your hand? What color are your shoes? Look in a mirror. What color is your hair? Now try the green plastic. What's going on?

The colored plastic is "filtering" the light. In other words, when you look through the colored plastic it lets only light of the same color reach your eyes, and filters out other colors.

16

Using crayons or markers, print your full name on a sheet of white paper, but make some letters red and other letters green. Look at your name through the red plastic. What happened? Now look through the green plastic. What did you see?

The red plastic filters out the red in everything you look at, and the green plastic filters out the green. When you look at the red letters through the red plastic, they seem to have disappeared. But when you look at the green letters through the red plastic, they look black.

GOING FURTHER

What would all of these things look like if seen through blue plastic, or orange, or purple? Start a colored plastic collection and find out!

Spinning Colors

If you mixed red, yellow, and blue together using paint or markers, you'd end up with a sort of brownish-black. But there's a way to mix these colors together and make a completely different color.

You will need:

a pencil
the plastic lid of a small yogurt container or
  take-out coffee cup
white paper
markers or paints in red, yellow, and blue
scissors
tape
round toothpicks (not the flat kind)

With the pencil, trace the circle of your plastic lid onto the white paper.

Divide the circle into three equal sections as shown, and color one red, one yellow, and one blue.

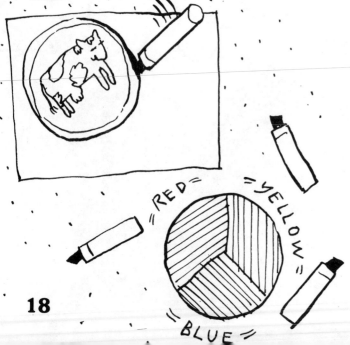

Cut out the colored circle with the scissors. Fold it in half, and then in half again. The point where the two folds meet is the center of the circle. Mark it with the pencil.

Tape the colored circle to the top of the plastic lid, then stick a toothpick through the center, as shown.

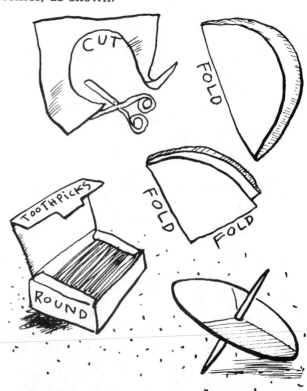

White Light

Sir Isaac Newton created *this* experiment, too. Having shown that he could separate sunlight into the colors of the spectrum, he wanted to show that he could put the spectrum back together again into "white" sunlight!

You now have a top. Give it a spin and keep a close eye on the colored circle. What color do you see as the top spins quickly?

Instead of brownish-black, you probably saw something more like creamy gray. If you could have used all the "true colors" of the spectrum, you would have seen the purest white.

The spinning colors do not actually mix — when they stop the colors are still separate on the wheel. But the spinner turns so fast that your eye combines the different colors of light that you see into a single color.

GOING FURTHER

Make more tops, each with a circle painted in a different color combination and different designs. What colors do you see when you spin these tops?

COLOR

What's your favorite color? Most people have a color they feel most comfortable with. Do you? Is it a "warm" color, like red, orange, or yellow, or is it a "cool" color, like blue, green, or purple?

There's more to color than meets the eye. We often use colorful expressions to talk about our feelings. Have you ever felt blue, green with envy, or in the pink? Although the world in black and white can be very dramatic, it's pretty hard to imagine what life would be like without color.

Have you ever wondered where colors come from? Colors are actually either animal, vegetable, or mineral. People have been using dyes made from the earth to color their clothes, hair, nails, and bodies since ancient times. Colors come in an amazing array. There are at least 27 words just to describe different types of red!

scarlet	rouge	tomato
crimson	cinnabar	strawberry
maroon	wine	raspberry
burgundy	vermilion	watermelon
magenta	Prussian red	cadmium red
rose	Venetian red	Tuscan red
mulberry	red ocher	Vandyke red
ruby	Thalo red	red oxide
garnet	rose	cherry

With a few simple ingredients you can make colors appear . . . and disappear . . .

Color Combos

Take a look around you. Did you know that all the colors you see — all the oranges, greens, purples, and browns — can be made from just three colors? By mixing red, yellow, and blue, you can make every color in the world!

You will need:

a white Styrofoam egg carton or a white
   plastic ice cube tray
a cup of water
food coloring (available at the supermarket in
   a package of four little squeeze bottles: red,
   yellow, green, and blue)
an eye dropper

Fill each compartment in your egg carton or ice cube tray half way with water. Add a couple of drops of the red food coloring to one of them. How red does the water become? Use the eyedropper to add more water and the red will become lighter. Add more food coloring and the red will become darker.

In another compartment try making dark and pale blues. Then mix a blue and a red together. What happens? It will make purple. Try creating a rich red-violet, the color of grape juice, or a deep blue-violet, like the skin of an eggplant.

Now try the yellow. Add it to the red and see what happens. Now mix yellow and blue. What colors did you get? Can you make an orange orange or a bright lime green? By mixing combinations of red, yellow, and blue, you can make any color you want. Can you mix a color that is exactly the color of something you're wearing? Suprisingly, black isn't always an easy color to make. Try mixing up a batch of black.

Color Terms

There are many words that people use to describe different colors. Here are a few:
   hue: a pure color of the spectrum
   tint: a hue mixed with white
   shade: a hue mixed with black
   tone: a hue mixed with gray

Taking Colors Apart

You've tried color mixing, now try color separating with a trick scientists call paper chromatography (kro-muh-TOG-ra-fee). It's a very simple process used to separate mixtures into their different colored chemical parts.

You will need:

a paper towel
scissors
water-based markers, including black
a glass
water
tape
a pencil

Cut the paper towel into strips about one inch wide.

Mark a strip of towel just above the bottom with a black marker, as shown.

Put an inch of water in the glass. Now tape the strip of towel to a pencil, as shown in the picture. Make sure that only the very tip of the towel touches the water, and not the spot that you made with the marker.

Watch closely as the paper towel absorbs the water. What do you think will happen when the water reaches the marker spot? As the water creeps up the towel and passes the spot, it takes some of the marker with it. What happens to the towel strip after ten minutes? An hour? A day?

The black in the marker is made of many different colored dyes. The different colored dyes travel up the towel at different speeds, leaving a multicolored pattern.

GOING FURTHER

You can try this paper chromatography test on lots of different drawing materials. Try separating the inks in several different brands of black, green, and brown markers. Watercolor paints, inks, and poster paints are fun to separate too.

Making Ink

Have you ever slid down a grassy hill only to get up and find your pants stained green? The grass is full of a strong green that is crushed out of the grass and into the fabric of your pants.

Here's a way to take the green out of grass and put it in a bottle of green grass ink.

You will need:

several handfuls of freshly-picked grass
scissors
an empty, rinsed-out tin can (with no sharp
   edges)
water
a wooden spoon
a clear glass
a paintbrush or Q-tip cotton swab
white paper

Use the scissors to cut the grass into little pieces, or tear it with your hands. Stuff the cut grass into the tin can. Add a little water and start mashing the grass with either end of the wooden spoon. With just a little action of your arm and wrist, you'll soon have green ink.

Pour some of the liquid into a clear glass to check its color. If you want to make it darker, pour it back into the can and mash the grass a little longer. If you want to make it lighter, add a little more water.

When your ink is the color you want, pour it into a glass, dip a paintbrush or Q-tip cotton swab into it, and make a drawing on bright white paper. Green grass ink is great for pictures of a forest of pine trees, a cave full of dragons, or, of course, a field of green grass.

GOING FURTHER

If you want to make other colored inks, try mashing up blueberries, dandelion flowers, or red beets. What plants do *you* think would make strong-colored inks?

Curious Colors

Today the colors in most inks and dyes are made from chemicals from a laboratory, but these "dyestuffs" used to come only from plants and animals. Here are some of the things people used to treasure for the color they could extract from them:

Plant or Animal	Color
the ink sac of the cuttlefish	brown
the leaves of the indigo plant	blue
the bloom of the safflower	yellow
the whole body of an insect called the cochineal	red
charred bones and horns	black
the body of the murex snail	purple

Invisible Messages

You can use kitchen chemicals to make invisible ink. When you send this kind of secret message to a friend, he or she will need some grape juice to decode it!

You will need:

two cups
a tablespoon
baking soda
water
a paintbrush or a couple of Q-tip cotton swabs
a sheet of white paper
purple grape juice

In a cup, dissolve one tablespoon of baking soda in one tablespoon of water. Believe it or not, this simple solution is your invisible ink.

With a paintbrush or Q-tip cotton swab full of this ink, write your name on a sheet of paper. Let the watery words dry completely. You now have what appears to be a blank sheet of paper. However, you know it's really your invisible signature.

Pour a little grape juice into another cup. Dip the paintbrush or a clean Q-tip cotton swab into it and paint over your invisible writing. Your name will mysteriously appear in vivid blue-green letters!

GOING FURTHER

To unlock the mystery of making invisible messages, read the next activity. Meanwhile, tell your best friend about this trick and have fun sending secret messages to each other.

Changing Colors

Here's a way to turn paper painted with cabbage juice from purple to bright pink or to startling blue-green!

You will need:

a fresh red cabbage
a bowl or pan
hot water
a paintbrush or a couple of Q-tip cotton swabs
several sheets of white paper
vinegar or juice from a lemon
three cups
a tablespoon
baking soda
water

First you need to make some cabbage juice. You can't squeeze it out of the cabbage, you need to "cook" it out. (You may need an adult to help you with this part.)

Cut the cabbage into small chunks and cover them with very hot water. Let the cabbage soak in the hot water for about twenty minutes. When it has cooled, throw out the cabbage pieces but save the reddish-purple water: this is your cabbage juice.

Paint a few sheets of paper with the cold cabbage juice, covering the entire sheet, and set them aside until they are completely dry. Then pour a little vinegar or lemon juice into a cup and dip a paintbrush or Q-tip cotton swab into it. Make a drawing on a sheet of juice-stained paper. Watch what happens to the purple paper as it comes in contact with the liquid. It turns from purple to pink!

In another cup dissolve one tablespoon of baking soda in one cup of water. Use this baking soda and water solution to paint a picture on another sheet of juice-stained paper. What color did the purple paper turn this time? A bright blue-green.

Acids and Bases

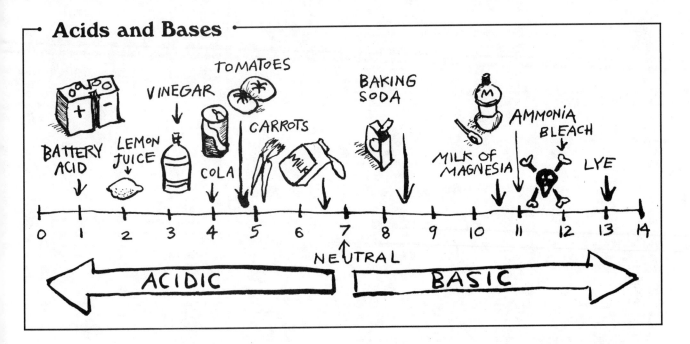

All of this is possible because red cabbage juice is a natural *indicator*. An indicator shows you something. Red cabbage shows you whether things are acids or bases or neutral. (An acid or a base is a chemical solution or compound with certain properties.) If something turns the red cabbage pink, it is an acid. If it turns the red cabbage blue or green, it is a base. If there is no color change, it is neutral.

There are many kinds of acids. You may have heard of citric acid, the stuff that makes lemons sour, or acetic acid, which is another name for vinegar. Can you think of some other kitchen liquids that might be acids? Try testing cola, coffee, or orange, grapefruit, or lime juice.

A base is just the opposite of an acid. Baking soda, laundry soap, and mouthwash are bases. The red cabbage indicator shows you a base by turning a blue-green color.

When acids and bases are mixed together they balance each other out and produce neutrals. So if you want to turn the acid pink of the lemon juice or vinegar back to the cabbage purple, you can neutralize it with a base like baking soda. To neutralize a base? The blue-green color can be changed back to a neutral purple cabbage juice color with just the right amount of an acid like lemon juice.

In the "Invisible Messages" activity, the purple grape juice was an indicator. When you painted it over the "invisible" baking soda it turned blue-green, revealing that the baking soda was a base.

SHADOWS

Have you ever been followed by your shadow? Or have you ever followed it?

Shadows, like rainbows, need the right conditions to appear. On cloudy days it can seem as if there are no shadows at all. Yet on a clear day, if you stand with your back to the sun, you are sure to see your shadow.

Light cannot go through your body or curve around it. When your body blocks sunlight, it makes a shadow. Other things make shadows too. Look around to see how many different shadows you can find. Does a clear glass make a shadow?

Shadows can help you tell time, draw a picture, or tell a story!

You and Your Shadow

Go outside on a sunny day and find your shadow on the ground. Does it look exactly like you? Does it move when you move?

Take your shadow for a walk. Is it following you or are you following it? Can it go everywhere that you go? Or does it sometimes disappear?

You can change the shape of your shadow. Try this by casting it on a wall, a car, some stairs, or the curb.

Two shadows can be more fun than one. Here are some shadow games to play outside with others.

You will need:

a friend (or two, or three, or more!)
a piece of chalk
a sunny day

See if you and a friend, without actually touching each other, can make your shadows touch. Can you make your shadows shake hands?

Using a piece of chalk, trace the shadow of a friend onto the pavement. Can you fit your own shadow into the outline?

With some more kids you can try making a three-, or four-, or even five-headed shadow monster. How many arms does your monster have?

With a group of kids you can play a game of shadow tag. The object is to tag your friends' shadows with your foot, or with your own shadow. You don't always have to run fast to keep your shadow out of trouble — sometimes you can hide it in a bigger shadow, such as that of a tree or building.

Sun, Shadows, and Time

Your shadow is sometimes bigger than you are, sometimes much smaller. When the sun is low in the sky, your shadow is long and tall. When the sun is overhead at noon, your shadow is short and wide. Noticing how your shadow changes as the day goes by can help you estimate the time. You can keep a shadow record of something that doesn't move around as much as you do.

You will need:

a piece of chalk
a sunny day

Early in the morning, go out on the sidewalk or driveway and look for the shadow of something large and stationary, like a tree or the edge of a building. Trace the shadow on the pavement with chalk. Trace it again after half an hour, and again an hour later. Did the shadow move? Check in on the shadow just before sundown. Is it where you thought it would be?

The Shadow Knows

Your shadow markings can give you an idea of how a sundial works. A sundial is a kind of clock used by people of ancient Greece, Rome, Egypt, and China. Unlike most modern clocks, a sundial has no moving parts — except, of course, the shadow moving across its face.

The piece of the sundial that casts the shadow is called a gnomon (NO-mun), which is Latin for "one who knows" (the time, that is!). The sundial's face has numbers, and you tell the time by looking at which number the shadow is "pointing" to.

Portrait of a Shadow

Asilhouette (sill-oh-WET) is a shadow portrait, a paper cutout of someone's shadow profile. The word comes from the name of an unpopular French Minister of Finance, Etienne de Silhouette (1709–1767), who happened to make shadow portraits as a hobby. Since people thought he was stingy, silhouette also used to refer to cheap things.

Before photography was invented, people went to silhouette studios to have portraits made; it was less expensive and took less time than sitting still for a painter. Today few people make silhouettes for a living, but lots of people still make them at home for fun.

You will need:

a friend
a chair
a lamp with its shade removed
tape or thumb tacks
one large sheet of white paper
a pencil
a black marker or black crayon

Hand Shadows

You can make shadow shapes with your hands and fingers. Can you make this bird? What other creatures can you make?

Put a chair close to a blank wall. Leave just enough room between the wall and the chair for you to pass between them. Turn the chair sideways, so that when your friend sits in it his or her cheek is facing the wall.

Place the lamp about eight feet away from the wall and at about the same height as your friend's head. Darken the room as much as possible by closing the door and drawing the shades.

Move the lamp closer or farther away from your friend's head until the shadow on the wall is sharp and clear. Can you focus the shadow so that the outline of even your friend's eyelash is clear?

Character Silhouettes

The French artist August Edouart (1769–1861) created very detailed silhouettes. Many of the shadow portraits he made were full length; he believed that shadow pictures of the head alone could not capture a person's true character.

He created his amazing silhouettes very quickly, clipping most of them in under three minutes! During his career, he cut portraits of such famous people as Henry Wadsworth Longfellow, King Charles X of France, and American presidents Martin Van Buren and William H. Harrison.

Once you have a sharp shadow profile, tack or tape the paper onto the wall so that the shadow falls on it. Now you can carefully trace the shadow outline onto the paper with a sharp pencil. Getting your friend to sit perfectly still will be the hardest part! It might help to give your friend something interesting to look at, or some music to listen to.

When you've finished drawing your shadow portrait, its outline can be colored in with black crayon or marker. Or you can cut out the drawing with scissors and mount it on a sheet of dark-colored paper.

Shadow Show

Shadow puppets have been used to tell stories in many countries.

Shadow puppets are held behind a cloth or paper screen, and a nearby source of light casts their shadows onto the screen. The audience sits in front of the screen and sees only the shadows of the puppets, moving almost magically. Create some magic at home by putting on your own shadow shows.

Before you make shadow puppets, you'll need to make a theater for their performance.

You will need:

an empty cardboard box
a piece of wax paper or white paper
tape
a lamp without its shade
scissors
an assortment of small household objects to
  make shadows with

Cut a large rectangular hole in the bottom of the cardboard box. On the outside of the box, tape a sheet of wax paper or white paper over the hole. This is your shadow screen.

Place a lamp behind the shadow screen, or put the screen in front of a sunlit window. Try holding a household object — scissors, for instance — behind the screen to see what kind of shadow it makes. Focus the shadow of the scissors on the screen so that someone looking at the outside of the box can see the shadow clearly.

Lots of things around the house will make good shadow puppets. And many familiar objects will make very strange shadows. Try holding up cookie cutters, a fork, or a toothbrush behind the screen and see what shadows they create. Challenge your friends to recognize each object only by its shadow. Can you fool them?

Moving Puppets

You can easily make a shadow puppet with movable parts.

You will need:

lightweight cardboard or oaktag
scissors
string
transparent tape
drinking straws

First draw the shape of your puppet on cardboard or oaktag. Then cut out the parts that you want to move. To make a joint, attach two pieces of cardboard (such as a leg and a body) by taping a short piece of string between them.

To hold your puppet, tape a straw "handle" to the main part of the figure. Then attach a straw to each moving part by tying a piece of string to the straw, then taping the other end of the string to the figure.

To practice moving your puppet, hold the straw that is taped to the main body with one hand and use your other hand to lift, turn, and wiggle the movable parts.

GOING FURTHER

Here are some magic effects you can achieve with shadow puppets:

Shadow puppets are generally held against the screen, but as a special effect the puppet can be moved back from the screen to make its silhouette less defined.

Disappearing acts and magical transformations can be done easily. Small shadows can be absorbed into larger ones to create these illusions.

© Diane Willow

REFLECTIONS

Mirrors are all around us. You can see yourself reflected in a window, doorknob, puddle, or an empty TV screen.

You can see yourself in these things because of the light they reflect back to your eyes. Only very smooth things give a clear, regular reflection. Water makes a good mirror when its surface is smooth. You can see yourself reflected in a puddle on a clear, sunny day. But if the wind blows across the surface of the puddle, the water becomes wavy and you can't see yourself anymore. That's because when light bounces off a rough surface it scatters in many different directions.

Glass has a very flat, smooth surface, but it doesn't always reflect well by itself. Most mirrors are made from a piece of glass coated on the back with a thin layer of silver, aluminum, or other reflective metal. This combination gives a bright and clear reflection.

In the activities in this chapter, you'll find that you can use mirrors to see around corners, write secret messages, or create ever-changing patterns. You'll also discover why what you see in a mirror is not the real thing.

⌐ Historic Mirrors ⌐

Mirrors are not modern inventions. In ancient times they were made of a thin disk of metal, generally bronze, polished on one side. In medieval times mirrors were made of steel and silver. Glass mirrors began to be produced in Venice in the 16th century.

Upside-down and Backward

What does the secret message below say? To find out, hold the page up to a mirror.

ɔigɒM ꙰oɿɿiM

You can read those backward letters in a mirror because the mirror reverses them. And with a little practice, you can write secret mirror messages yourself!

You will need:

a mirror
paper
something to write with

Set the mirror on a table as shown. Looking only at the reflection of what you are writing, try printing your name so that it reads correctly in the mirror. It's easy to get mixed up writing this way! You have to make all your letters upside-down and backward. Were there any letters you didn't have to write differently at all?

GOING FURTHER

Once you become an expert, try this trick on your friends and family. They'll be impressed by your skill.

Jabberwocky

In Lewis Carroll's *Through the Looking Glass*, when Alice passes through her living room mirror into the Looking-glass House, she finds all the books there are printed backward, like this:

'Twas brillig, and the slithy toves
Did gyre and gimble in the wabe:
All mimsy were the borogoves,
And the mome raths outgrabe.

This poem doesn't make much sense, even when read in a mirror!

Follow that Wriggle!

Can you trace this wriggly line with a pencil and not lift the pencil point from the paper?

Simple, you say? Of course it is — if you do it while looking at the line. But it's quite another thing to do it while looking only at the *mirror image* of the line instead of at the line itself!

You will need:

a mirror
a sheet of paper
a pencil

Stand the mirror in front of you on a table as shown.

At the top of your sheet of paper, draw a crazy, curvy line like the one above. Fold up the bottom half of the sheet so that you can't see the line you've drawn, only its reflection in the mirror.

Looking *only* in the mirror, put your pencil point on one end of the line and start to trace it. Remember, you can look only at the mirror image of the line, not at the line itself.

How could something that sounded so easy turn out to be so hard? It's because what you see in the mirror is the reverse of what you've drawn on the paper. Whenever you try to turn a corner with your pencil, your brain tells you

to turn the opposite way. You're being fooled by habit, and by the mirror image! It will take a little practice, but you'll soon learn a new way of thinking, and be ready to follow the trickiest, wriggliest line.

Secret Writing

Mirror writing has been used for centuries to discourage snoops. Leonardo da Vinci, the Italian artist and inventor, kept his secret notebooks in mirror code. The English novelist Jane Austen scribbled all her notes backward; when people tried to read them over her shoulder, they couldn't understand a word!

Kaleidoscope

Kaleidoscopes have been popular toys in America for more than a hundred years. They are usually in the form of a round tube with a peephole in one end. When you look through the peephole and rotate the tube, you see an ever-changing variety of colored patterns. If you have ever looked into a kaleidoscope, you've probably wondered what makes these incredible designs. The secret is mirrors and their reflections.

Here's a kaleidoscope that's fairly easy to make. It will take you about an hour to put together, and you may want to ask an adult to help you with some of the trickier parts.

You will need:

a rinsed-out quart-size milk carton
scissors
mirrored contact paper, which can be bought
  in a roll at a hardware store
masking tape
plastic wrap
two rubber bands
little pieces of ribbon in different colors
wax paper or tracing paper

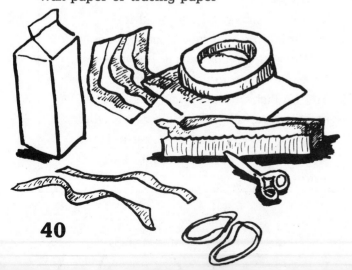

Cut apart the milk carton so that you have three connected sides, like so:

Using the three-sided milk carton as a pattern, cut out a piece of the mirrored contact paper and stick it to the milk carton. One side of your milk carton should be shiny now.

Fold your milk carton into a triangular tube shape, with the shiny part on the inside.

Tape the milk carton together.

Now look through the tube and see how a kaleidoscope works! The reflections from the mirrors make symmetrical patterns out of everything you look at through the tube.

Stretch a piece of plastic wrap over one end of your tube and fasten it in place with a rubber band.

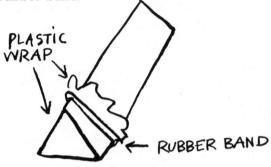

PLASTIC WRAP

← RUBBER BAND

Cut your ribbons into tiny pieces, about this size:

WAX PAPER OR TRACING PAPER

PLASTIC WRAP

Put the little ribbon pieces *on top* of the plastic-wrapped end of your tube.

Carefully place a piece of wax paper or tracing paper over the same end, and fasten it with a rubber band. The pieces of ribbon are now sandwiched between the plastic wrap and the wax paper.

Look through the open end of your tube. Shake your tube to move the pieces of ribbon and change the pattern you see.

Each time you shake the tube, the pieces of ribbon will create a new pattern. Do you ever see the same pattern twice?

Around the Corner, Over the Wall

Mirrors allow you to see things that you can't see otherwise. Rearview mirrors in cars let you see what's behind you without turning your head. Three-way mirrors in clothing stores let you see yourself, front and back. A periscope is a special device that lets you see around corners and over walls. You can make your own periscope and use it to see over a crowd at a parade or a performance.

It will take you about an hour to make this periscope. You may need to ask an adult to help you with some parts.

You will need:

two rinsed-out quart-size milk cartons
scissors
masking tape
a ruler
a pen or pencil
two small pocket mirrors, no smaller than
    three inches by three inches

Unfold the tops of the milk cartons and cut them off. Now cut the bottoms, too. Tape the two milk cartons together to form a long tube.

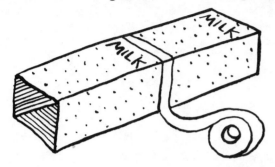

Cut a hole at the bottom of one side of the tube. Cut another at the opposite end of the other side of the tube.

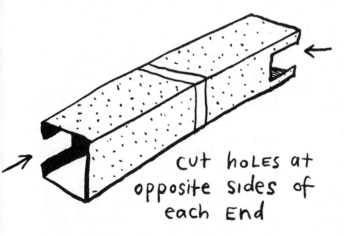

cut holes at opposite sides of each End

Special Periscopes

You've probably seen a periscope used in a submarine to see above water. Other specially designed periscopes can even show what's inside our bodies.

Using the drawing as your guide, cut slots in opposite sides of each end of the milk carton tube. These slots must be cut at 45-degree angles.

To find this angle, measure the length of the bottom edge of the milk carton. Then, starting at the bottom of the carton, measure that same length up one side, along an edge. Mark the carton, then draw a line from the mark to the far corner of the bottom edge. You'll have your 45-degree angle.

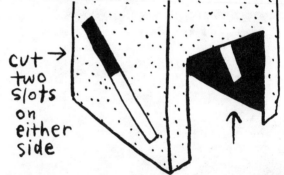

cut → two slots on either side

Slide your mirrors into the slots. The shiny sides need to face *into* the tube. Tape the edges of the mirrors to the outside of the tube.

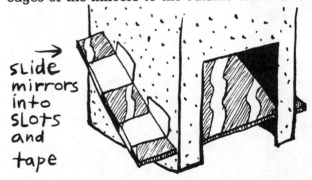

slide mirrors into slots and tape

Your periscope is now ready to use. When you look into the bottom mirror, you'll see whatever is reflected in the top one!

43

44

WATER

Water is in your body and covers most of the surface of the earth. You use it every day. In fact, in one day Americans produce 88 million gallons of moisture by breathing and perspiring.

Water fits into all kinds of containers. It may seem as if water doesn't have a shape of its own at all. But if you've ever watched the rain fall against a window pane, then you've probably seen the shape the water makes — a round drop shape. The water's surface has a tension that holds it together in small droplets.

Some very surprising things can happen in water. Have you ever seen coins sparkling at the bottom of a fountain or pool where people have thrown them in for good luck? The tiny metal coins sink right to the bottom. So how can a huge metal ocean liner float? Whether something floats or sinks depends on a lot of things — its shape, weight, and the air it holds.

Sink . . . or Float?

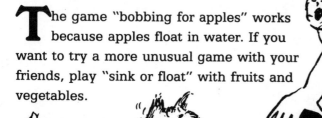

The game "bobbing for apples" works because apples float in water. If you want to try a more unusual game with your friends, play "sink or float" with fruits and vegetables.

You will need:

a dishpan, bucket, or bathtub full of water
old newspapers
fruits and vegetables, whatever you have on
   hand — try carrots, potatoes, onions,
   bananas, lemons, limes, olives, string beans,
   or some special fruits such as melons or
   coconuts

Find a place to set up this activity — outdoors is great. If you do it inside, be sure to spread lots of newspaper around your container of water to catch any splashes. Line up your collection of fruits and vegetables beside your tub of water. Pick any fruit or vegetable and hold it in your hand. What is it like? Is it heavy, or is it light? Is it solid, or hollow? Does it have holes in it?

Ask your friend to predict what your fruit or vegetable will do: sink or float? Place it gently on the surface of the water and test your friend's prediction. What happened? The expected or the unexpected?

You'll probably find that some of your fruits and vegetables behave strangely. Many that you thought would sink right to the bottom probably rose to the surface. Some will sink, but not down to the bottom of the tub.

GOING FURTHER

After testing everything, try cutting open the fruits and vegetables to see if you can discover the secret of why they sink or float. A clue: look for spaces that hold air.

A Floating Rock?

When the bubbly froth on volcanic lava hardens, it becomes a lightweight rock called pumice (pronounced PUM-is). You can buy a piece of pumice at the drugstore; many people use it in the bathtub to scrub the rough skin on their elbows and feet. Although it's a rock, it won't sink! Why? Because it's full of trapped air bubbles, which make it float.

Rafts and Barges

A raft is a flat floating platform. A barge is a sort of gigantic raft, a large, flat-bottomed boat designed to carry heavy cargo on canals and other calm waterways.

You can make a model raft or barge out of aluminum foil and create cargo for it.

You will need:

a five-inch square of aluminum foil
a dishpan, bucket, or bathtub full of water
"cargo," such as paper clips (at least 100!)

Carefully place the square of aluminum foil on the surface of the water. Will it still float if you put a paper clip on it? How many paper clips do you think the raft can carry before it sinks? Test your prediction. You'll probably be surprised at how many it takes.

When the raft sinks, dry it off and try again. Your raft will hold more paper clips if you add each one to the load s-l-o-w-l-y and g-e-n-t-l-y? Try spacing them evenly over the raft to see if it can hold even more.

GOING FURTHER

You and a friend can take turns adding paper clips to a raft until it sinks. Try not to be the one who sinks the raft!

You can try different kinds of cargo, such as: nails, marbles, pencils, cotton balls, toothpicks, dried beans, or peas.

Buoyant Boats

It might seem that all heavy objects sink and only light ones float. But you can float things that usually sink — if you float them in a boat.

You will need:

Plasticine clay (the kind that doesn't dry out)
a dishpan, bucket, or bathtub full of water
"cargo," such as paper clips or marbles

Roll a lump of Plasticine into a sphere about the size of a golf ball. Drop it into the water and it will sink. But shape it into a bowl-shaped boat and it will float. Don't be discouraged if your first boat goes to the bottom. Keep shaping the Plasticine until you make one that floats. (Hint: the one that will most likely float will look like a wide, deep bowl with high sides.)

Load your boat with paper clips, marbles, pebbles, or any other small objects you can find. As you load the boat, it will sink lower and lower in the water. If the boat pushes down as hard as the water pushes it up, it will float. This is called buoyancy. If the boat pushes down harder than the water pushes it up, the boat will sink.

GOING FURTHER

See if you can make clay boats of other shapes that float — try rafts, canoes, and rowboats.

Hydrometer

Y ou've probably heard that it's easier to float in salt water than in fresh water. But is it true? Find out for yourself with the right tool for the job, a homemade hydrometer (hi-DRAH-muh-ter).

You will need:

two identical, tall, clear glasses of water
⅓ cup of salt
a spoon
a plastic straw
scissors
Plasticine clay (the kind that doesn't dry out)
thread

Set the two glasses of water on a table and add the salt to one of them. Stir the salt water, then let it stand while you make your hydrometer.

Cut the straw so it's just a little shorter than the glasses you're using. Roll a chunk of Plasticine into a small ball about the size of a marble and stick it firmly on the end of your straw. Now try to float the straw, clay end down, in the glass of fresh water. You may have to add or remove some clay to get it to balance like it does in the picture. Be sure that the connection between the clay and the straw is tight or the straw will fill with water and sink.

When the straw floats well, take it out of the water and tie a piece of thread around its middle. Float it in the water again and slide the thread up or down the straw until it marks the water level. You have made a hydrometer.

Now float your hydrometer in the glass of salt water. Where is the thread? In which does your hydrometer float higher, the fresh water or the salt water?

Your hydrometer measures the density of liquids. The more dense a liquid is, the higher your hydrometer will float. Which liquid was more dense?

GOING FURTHER

Are there other liquids you can test with your hydrometer? Try fruit juice, milk, oil, or vinegar.

The Skin of a Water Drop

The surface of water is like a thin skin. This skin is held together by a force called surface tension. With this experiment, you can actually *see* the skin of the water!

You will need:

a glass of water
a paper towel
an eyedropper

Place the glass on top of the paper towel and fill it up to the very top with water. Use your eyedropper to add even more water to the glass. How many drops can you add before a bead of water trickles down the side of the glass and onto the paper towel?

Now get down to eye level with the top of the glass. What do you see? That bulge of water is being held together by the water's surface tension.

Walking on Water

Water striders are insects that can actually walk on water. They live on the surface of ponds and other still bodies of water. The water strider can walk on water because it carefully avoids breaking the pond's surface tension. It has long legs to spread its weight over a large area, and it has waterproof hairs to keep itself dry.

GOING FURTHER

Take turns with a friend adding water to a full glass — and try not to be the one who finally makes it overflow!

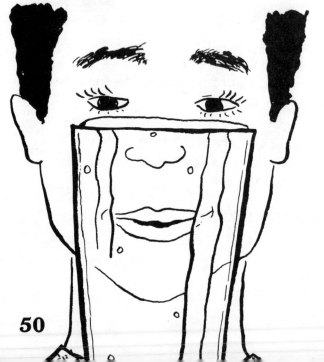

Soap Power

You may have seen motorboats before, but probably not like this — an aluminum "motorboat" powered by dish soap!

You will need:

aluminum foil

dishwashing liquid

a large dishpan or bucket filled with an inch of water

Shape a small boat out of aluminum foil. A triangular boat like the one in the picture will work especially well.

Squirt a blob of dishwashing liquid onto the outside of the boat's back end. Make sure that the blob is on the boat. It must make contact with the water.

Put your boat gently in the water and watch it go!

Water is more attracted to soap than to itself. The soap breaks the surface tension of the water and so acts as a natural engine. Once you've completely broken the surface tension, however, you can't put it back together again; if you want to send your boat on a second trip, you'll have to change the water and start over.

Breaking the Tension

When you put some drops of water onto a piece of wax paper they take the shape of domes. You can flatten the roundness of water droplets with a touch from a finger that is coated with dishwashing liquid. One touch from the soap flattens the water drop.

WIND

There are no photographs of the wind, but you've probably seen pictures of the effects of the wind — whether it's a powerful, destructive tornado or a steady breeze filling the sails of a boat.

Although the wind is invisible, you know when it's around you. The force of a strong wind can pop an umbrella inside out. At other times the air seems very still — even clouds high in the sky don't seem to move.

Get out a pencil and jot down as many names of the wind as you can think of. How many did you come up with? You probably wrote down some common ones, such as breeze, hurricane, tornado, and gale. But did you know that there are hundreds of names for different kinds of winds?

It's a breeze to find out more about the unseen force around you. Just try the following activities!

World Winds

chinook: hot, dry wind of the U.S. Rocky Mountains. This wind produces warm temperatures and can quickly melt snow.

doldrums: calm winds near the equator

fohn: hot, dry European wind that flows down the mountainsides of the Alps

haboob: a violent wind that raises sandstorms which move across the land as a bright yellow wall and are followed by rain. Haboob comes from the Arabic word *habb,* meaning to blow.

mistral: a cold, dry, blustery northwest wind that blows along the coasts of Spain and France. The name comes from the Latin word *magistralis,* meaning masterly.

nor'easter: a cold wind in the New England region of the U.S.

sirocco: a hot, dry southeasterly wind. It's usually a spring wind that blows across North Africa from the Sahara.

waff: gentle breeze in Scotland

zephyr: mild breeze bringing pleasant warm weather

Streamers

Here's a way to find out which way the wind is blowing by turning yourself into a crazy sort of walking weather vane!

You will need:

one or many of the following lightweight
    materials: tissue paper, wrapping paper, old
    newspaper, ribbons, crepe paper, plastic
    lawn bags
scissors
masking tape or transparent tape
lightweight cardboard or oaktag
a stapler

Cut your lightweight materials into strips about two inches wide and two feet long. Now tape some of these strips to your wrists and ankles and tape the rest to your arms and legs. When you've covered yourself with streamers, go outside to explore the wind. Your neighbors will wonder what in the world you're up to! If you want to add to your costume, staple a strip of light cardboard or oaktag into a ring and attach streamers to it to wear as a hat, bracelet, or anklet.

Your streamers will tell you which direction the wind is coming from. Take a walk around the neighborhood. Look for places where the wind doesn't blow at all, and places where your streamers wave wildly around you.

Bubble Dance

Soap bubbles are as light as the air and are wafted along by the slightest breeze. By following the adventures of a floating bubble, you can see the path of the air current it's riding.

You will need:

dishwashing liquid (Joy and Dawn seem to make the strongest bubbles)

a bucket for water

warm water

a variety of bubble makers, which can be almost anything with holes: an empty can with both ends cut out of it, the plastic holder from a six-pack of soda, or the plastic container with little square holes that berries are sold in. You can also use your hands to make different shaped openings.

You can easily make your own bubble solution. Fill your bucket with two cups of warm water: it feels nicer than cold water to dip your hands into, and it will make the bubbles last longer. Add ¼ cup of dishwashing liquid to the water and take your bucket and collection of bubble makers outside.

Dip a bubble maker into the bubble bucket until you get a soap film across its hole or holes. Wave your bubble maker in the air or blow gently into the film. You'll soon fill the air with beautiful soap bubbles.

Make one big bubble and follow it with your eyes. How high does it go? Does it stay near you or move away fast? Does it soar into the air or hover near the ground? Does it travel a straight course or a crazy one?

If you have a plastic berry container, make a flock of little bubbles and watch them fly. Do they travel together in a pack or do they dance away from each other in all directions?

Go Fly a Kite

If you could fly, you would feel the winds that move the clouds. You may not be able to touch those high winds, but one way you can get a sense of them is to fly a kite.

This kite is easy to make and *always* flies. It's called a sled kite. The design for the sled kite comes from experiments done to improve parachutes.

It will take at least an hour to make this kite. If you can, get a friend to help you; that will make it easier and much more fun.

You will need:

one large extra-strong trash bag, 24 inches by
   30 inches (a sheet of plastic this size, such
   as the kind you put over windows to keep
   the cold out, will also work well)
a yardstick
a marker
scissors
two wooden dowels, 24 inches long and ⅛
   inch in diameter (you can get these at any
   hardware store; tell the clerk exactly what
   you want)
duct tape
a hole punch
a new ball of string

Clear a space on the floor or find a large tabletop where you can spread out your plastic perfectly flat. We'll call the longer edges the "top" and the "bottom," and the shorter edges the "sides."

On the bottom edge of the plastic, measure in six inches from one side and mark the plastic with a marker, then make another mark six inches from the other side. Make the same marks on the top edge of the plastic. Now, along the edge on each side, measure down five inches from the top and make a mark. Use the yardstick to draw lines with the marker connecting the six marks you've made, and you'll have the six-sided shape shown in the picture.

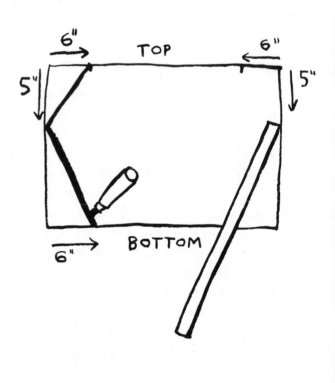

Cut out your six-sided piece of plastic with scissors. Lay the dowels down on the shape, from top to bottom, as shown in the picture. Tape the dowels securely to the plastic with duct tape, making sure you tape down the top and bottom of the dowels and not just the middle.

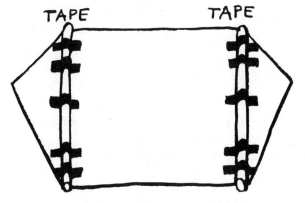

Cover the front and back of the two outside corners — the ones not connected by the dowels — with duct tape. Punch a hole in each taped corner with the hole punch. Now your kite is ready for kite strings.

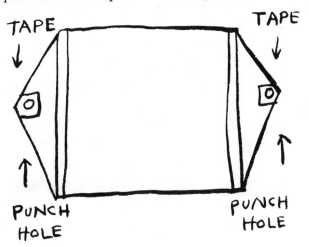

Measure and cut a three-foot length of string. Fold the string exactly in half and mark the fold with a marker. With your forefinger on the mark, hold the string down on a tabletop and have a friend tie a knot with the string around your finger. When you take your finger out, there should be a loop in the middle of the string, just like the one in the picture.

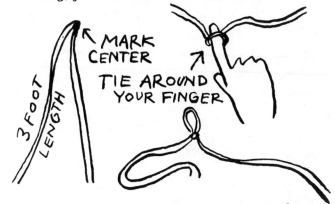

Tie the two ends of the looped string to the holes you made at the corners of your kite. This string is called the bridle string. Now take the loose end of the ball of string and tie it securely to the loop in the bridle string.

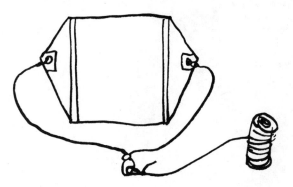

Your kite is ready to fly! The next activity will help you become an ace kite flier.

Up, Up, and Away!

Kites and spring days seem to go together, but you can fly your kite in any season, as long as it's not raining or snowing. Choose a sunny day when steady winds are blowing at between four and fourteen miles per hour. That's the ideal wind speed for kite flying. The Beaufort Scale on the opposite page will help you figure out the wind speed.

Find a safe open place to fly your kite. "Safe" means a place with no trees, buildings, traffic, telephone wires, or power lines. Playgrounds, parks, fields, and beaches are ideal.

To launch your kite, stand with your back to the wind. In one hand, hold your kite in front of you by the bridle string; in the other hand, hold your ball of string. Soon you will feel the wind pushing against the front of your kite, urging you to let go. Let out a little string from your ball and pull gently as the wind lifts your kite into the air.

Gradually let out more and more string, pulling a bit as you do. Keep doing this until your kite is launched into the air. If it begins to fall, give the string a tug until it begins to rise again. If it falls to the ground, don't get discouraged. Soon you'll get the hang of it and be able to keep the kite airborne for any length of time.

To bring your kite back to earth, reel in the string slowly and carefully. As the kite nears the ground, walk toward it as you reel it in. Soon it will be ten yards away, then five, then one . . . then at last you can take it home to fly another day.

Watching the Wind

Sir Francis Beaufort (BOH-for) (1774–1857) was a careful observer of the wind. He developed a system, still used today, for measuring and classifying wind speeds.

Once you learn the 13 speeds on the Beaufort scale, you'll soon be able to tell how hard the wind is blowing just by looking out the window.

0	calm	less than 1 mile per hour	smoke rises straight up
1	light air	1 to 4 miles per hour	smoke slowly drifts, but weather vanes do not turn
2	light breeze	4 to 8 miles per hour	you feel the wind on your face; leaves rustle; weather vanes turn
3	gentle breeze	8 to 12 miles per hour	the wind keeps leaves and twigs in constant motion; flags flap
4	moderate breeze	12 to 18 miles per hour	the wind keeps small branches in motion and raises dust and loose paper; mosquitoes stop biting
5	fresh breeze	18 to 24 miles per hour	small leafy trees sway; little white wavelets form on lakes and ponds
6	strong breeze	24 to 32 miles per hour	large tree branches move; telephone wires whistle; umbrellas are difficult to use
7	moderate gale	32 to 38 miles per hour	whole trees in motion; walking against the wind becomes difficult
8	fresh gale	38 to 48 miles per hour	the wind snaps twigs off trees; walking against the wind becomes very difficult
9	strong gale	48 to 54 miles per hour	the wind snaps branches off trees; shingles fly off roofs; children are blown over
10	whole gale	56 to 64 miles per hour	whole trees uprooted; adults are blown over; buildings are considerably damaged
11	storm	64 to 72 miles per hour	seldom experienced inland, but if it is . . .
12	hurricane	more than 72 miles per hour	. . . look out!

BALANCE

Balancing is a part of so many daily things that it seems to happen automatically. If you've ever swung on a swing or ridden a bike, then you've used your sense of balance.

To check your sense of balance, stand with your feet together and your arms at your side. Do you feel steady and balanced? Great! Now lift your left leg and stretch it as far as you can to the left. You'll feel your weight shift. You may lean over or even lift an arm up to keep your balance.

Each time you walk you shift your center of gravity. As you take a step, your center of gravity moves forward. You lose your balance just for a split second as you shift your weight onto your forward foot to balance yourself again.

Balancing Toys

Surprise yourself and your friends by making this incredible toy. No matter how far you tip it, it always regains its balance. The secret is keeping the center of gravity low.

You will need:

two sharpened pencils
the cork from a wine bottle
a round toothpick (not the flat kind)
paper
paints, markers, or crayons
scissors
tape

Stick the points of two pencils into opposite sides of the cork, as shown in the picture.

Now try to balance this contraption on the tip of your forefinger. You may need to adjust the pencils a few times to achieve perfect balance.

Now stick a toothpick deep into the center of the end of the cork. Place the whole thing on your fingertip. The weight of the pencils and the cork is now balanced on the tiny pointed end of a toothpick! With a little practice, you'll be able to spin your toy on the very end of your pinkie — and still keep it balanced!

GOING FURTHER

Turn this balancing trick into a toy by adding two colorful creatures to the ends of the pencils. Use paper, markers, and your imagination to draw things that will seem to fly or float as it spins around. They could be swimming fish, flying birds, drifting clouds, or anything else you come up with. Cut them out and tape one on each of the pencil ends. Balance your toy on the top of a soda bottle and give it a whirl!

Water Maze

A maze is any kind of complex and confusing path. Although some mazes are easier to find your way through than others, most are designed to tease your brain. You can make a water maze that will challenge your balancing skills as well as your mind. By tipping the maze from side to side you can move a waterdrop through its paths. With the help of gravity and a keen sense of balance, you'll know how to make the next move.

You will need:

a sheet of heavy white drawing paper
a pen or pencil
a slightly larger sheet of wax paper
tape
a glass of water

Draw a crooked, winding maze on the sheet of drawing paper. It can be as simple or as crazily complicated as you want, but it must have a definite beginning and end.

Cover the drawing with a sheet of wax paper, folding the wax paper over the edges of the paper and taping it securely to the back of the drawing. Your maze is now waterproof!

Place a single drop of water at the beginning of your maze. Tip the paper back and forth to make the drop roll down the path you drew. Try to get the water drop to follow the path of the maze all the way to the end. Which rolls faster, a big or a little water drop? Which is easier to control?

GOING FURTHER

You can make traps for the water drop by taping bits of paper towel to the maze. If a water drop hits the paper towel, it is absorbed! Create a water maze tournament with a friend and see who wins.

63

Mobiles

A mobile is a balanced, moving sculpture. The balance is so delicate that the mobile is moved by small breezes and air currents. Mobiles can be any size — huge metal sculptures in a park or small seashells dangling in a window. This simple mobile is an introduction to the fine art of balancing.

You will need:

string
scissors
3 plastic drinking straws
tape
6 paper clips
4 clean plastic lids in various sizes (from
    yogurt, milk jugs, cottage cheese, etc.)
crayons or markers
paper

Cut a piece of string about twelve inches long and tie one end of it tightly around the middle of a straw. Tape the string to the straw to keep it from sliding. Tape the other end of the string to a table edge or shelf so that the straw dangles freely in the air and you can work on your mobile easily.

Slip a paper clip over one end of the straw as shown. Oops! The straw tips up at the other end! Now add a paper clip to that end and slide the paper clip along the straw until the straw balances itself perfectly.

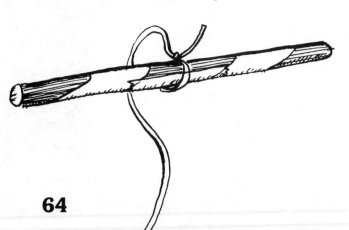

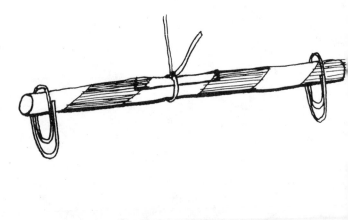

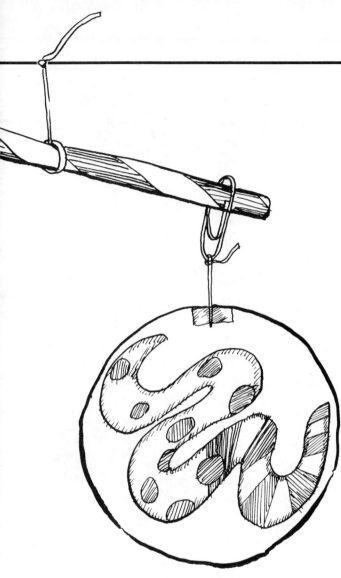

Tie the other string to the other clip. Stand back and check your balance. If one shape is heavier and tilts the straw, adjust it by moving its paper clip toward the center of the straw until the straw balances. When the shapes are balanced, tape the paper clips to the straw so they won't slip off. Now take the second straw and two more paper clips and follow the same steps to attach your other two shapes.

Choose two of your shapes to balance — they don't need to be the same size. Cut two pieces of string from which to hang these shapes on the mobile. Tape a piece of string to the top of each shape. Tie the free end of one string to one of the paper clips on the straw.

You can now join the two mini mobiles to create a bigger one. Tape the top string of each mini mobile to opposite ends of the third straw. If you make one of the strings shorter than the other one, all of your shapes will be able to move without becoming tangled.

Now you are ready to hang your mobile in a place where it will twirl in a gentle breeze.

GOING FURTHER

Try making a balancing mobile out of items you've collected: seashells, sticks, ticket stubs, cards.

Now that you've created a base for your mobile, you can decorate the lids that will hang from it. You can turn them into planets, eyes, insects, flowers, seashells, or anything else you can think of. Remember, your mobile will hang from the ceiling and the shapes dangling from it will seem to float in the air. You can tape on paper, ribbons, aluminum foil, feathers, or add stickers. Decorate both sides of the lids.

Salt Pendulum

A pendulum is a weight attached to a line so that it can swing freely back and forth. Here is an unusual kind of pendulum, one that records its own travels through space. The weight is a cup full of salt, and when it swings, it pours!

You will need:

a sharpened pencil
a paper cup
a ball of string
scissors
two matching chairs
a broomstick
masking tape
newspaper
a large sheet of black paper or cardboard
a box of salt

With the tip of a sharpened pencil, punch a small hole in the center of the bottom of a paper cup. Punch the hole from the inside of the cup to make it smooth and clean.

With a hole punch, make three evenly-spaced holes in the rim of the cup. Cut three pieces of string, thread them through the holes, and tie the three ends into a knot as shown in the picture above.

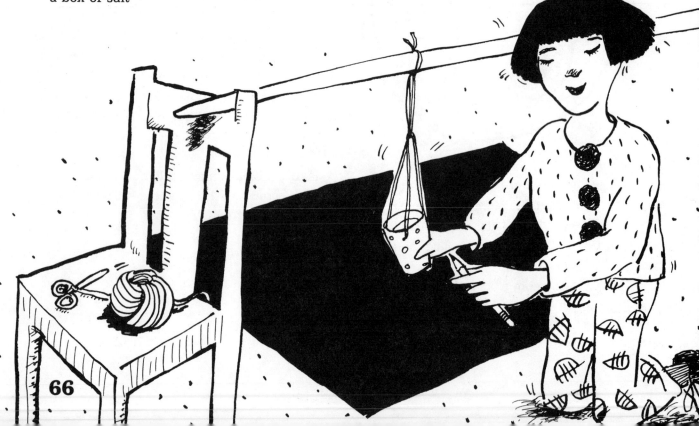

Set up two matching chairs back-to-back and rest a broomstick between the top of both chair backs. Tape the broomstick to the chair backs to keep it from rolling.

Cut a long piece of string, long enough to reach from the broomstick to the floor. Tie one end of the string to the knot in the paper cup string, as shown in the picture below. Tie the other end around the broomstick so that the paper cup, when dangling from the end of the string, hangs about an inch above the floor. You now have a free-swinging pendulum.

Cover the floor with newspaper to catch any spills. Lay a sheet of black paper on top of the newspaper beneath the cup. Cover the hole in the bottom of the cup with one finger and fill the cup with salt. Take your finger off the hole and start your pendulum swinging back and forth. As the salt pours out of the bottom of the cup, it leaves a beautiful white record of the pendulum's movements upon the black paper.

GOING FURTHER

What can you do to control the shape of the pattern? Can you make salt circles? X's? Five-pointed stars? Different ways of attaching the string at either end will create different patterns. Try the ideas shown below.

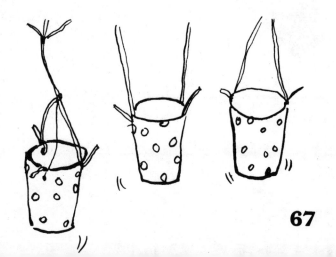

ILLUSIONS

What's real and what's not? Can you believe what you see? Not always, especially when you're dealing with the illusions found in the strange world of depth and distance.

Look at the photograph. Which do you think are larger, the windows near the bottom or top of the building? You probably know that in real life, these windows are the same size. But take a ruler and measure the upper and lower ones as they appear on the page — the lower ones are a lot bigger!

The bottom windows look bigger in this photograph because they're closer. Close objects look bigger; distant objects look smaller.

What about "dimension"? The photograph is completely flat — run your hand over it. But in real life the buildings are three-dimensional; that means you can walk around them.

Depth, distance, and dimension can all play tricks on your brain. With the next activities you can play games with your mind — and learn how to fool your friends and family.

Double Vision

Have you ever wondered why we have two eyes instead of one? This simple little trick will give you some idea.

Hold up one finger about five inches in front of your nose. Close one eye and look at the finger. Now look at the finger with only the other eye. Did the finger move? No, it's just that each of your eyes sees a slightly different picture of the world. When you look at the world through both eyes, your brain blends these two pictures into a single image.

You can play tricks on your brain with the different views that your two eyes see.

Another trick is called "The Hole in the Hand." Hold a cardboard or paper tube in your left hand. Put the tube right up to your left eye and look straight into it. Turn your right palm toward yourself and hold it against the side of the tube. Look! There's a hole in the middle of your right hand! A hole through which you can see the rest of the world! If you're having trouble seeing the hole, try switching eyes.

You will need:

your eyes, your brain, and your fingers
the cardboard tube from a roll of paper towels
   or a sheet of paper rolled into a tube

Here's a trick that fools the brain every time. It's called "The Floating Hot Dog." Hold your two forefingers tip to tip about two inches in front of your eyes. Focus your eyes on an object in the distance. Soon you will see a hot dog floating between your two fingers.

GOING FURTHER

It's hard to judge depth and distance if you use only one eye. Try playing catch with one eye open. (Use a soft ball so you don't get hurt!)

Mind Games

Your brain is used to figuring out what is near to you or far away by reading certain "clues" with your eyes. Every so often, though, your brain gets mixed up and misreads these clues. Here are some drawings guaranteed to confuse your sense of depth and distance.

Are the stairs right side up or upside down?

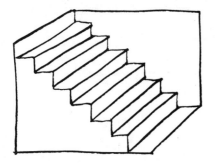

Which is the front side of this cube?

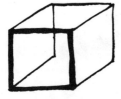

This?

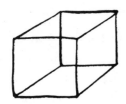

Or this?

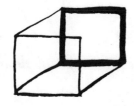

Try drawing your own cube. (A cube is a 3-dimensional square.) Here's how:

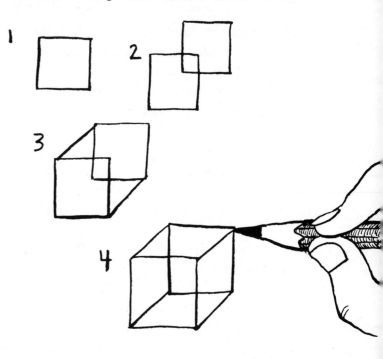

The changeable cube is an "optical illusion." (An *optical illusion* is something you think you see that is not really there.) As you look at the cube, it seems to change direction. But if you add shading and erase a couple of lines, the cube will stay in one position.

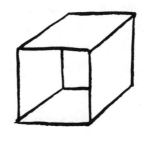
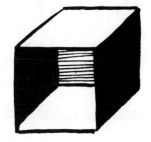

Diorama

A diorama is a lifelike miniature scene. It includes pictures and scenery set up in a box with an opening in one end. Dioramas usually create the illusion of great depth, so that when you look through the opening you feel as if you could step inside.

With simple household items and a little patience, you too can make a diorama.

You will need:

a shoebox with its lid
scissors
cardboard, construction paper, old magazines, small toys, etc.
glue or transparent tape
a lot of imagination

There's only one thing that every diorama must have, and that is a peephole. With the scissors, cut a half-inch hole into the middle of one of the short sides of the shoebox. This can be tricky; you might want to ask an adult for help.

What do you want to show in your diorama? It can be an indoor or an outdoor scene — one that comes from your imagination or something that you've seen. A beach scene with sunbathers, tiny sailboats and a white-hot sun off in the distance? A deep green jungle with gorillas peeking from behind giant ferns and snakes draped from branches? How about a diorama of your bedroom?

Once you've decided you can begin by drawing a picture or finding one in a magazine to make a background scene. Put this scene on the inside of the shoebox, on the end opposite the peephole.

Then create cardboard shapes to put inside the box, such as trees, buildings, waves, arches, furniture, etc. As you start putting together your diorama, peer into your peephole from time to time to see how things are shaping up. Before you start taping things down, decide if you want to decorate the bottom and sides of the inside of your shoebox to add to the environment. Do you have any paper dolls or toy figures that would make good "actors" in your diorama? Let your imagination run wild! Your finished scene could include almost anything: figures cut out of magazines and mounted on cardboard, rubber bugs, construction paper animals, twigs, pebbles, or shells.

Your diorama isn't complete until you've decorated the inside of the lid to your shoebox. Should there be a blue sky dotted with seagulls above your beach? Monkeys and vines hanging down into your jungle? A light hanging from your bedroom ceiling?

You can cut slits and holes in the lid of your shoebox to let light inside. Cover the holes with wax paper or colored tissue to create special lighting effects.

Keep experimenting! And keep your eyes open! Everywhere you look you'll find exciting new ideas for making dioramas. Window displays in stores and exhibits in old-fashioned historical museums can give you great ideas for dioramas.

Amazing 3-D

3-D glasses were a popular fad in the 1950s. If you wore the glasses while looking at a specially printed 3-D movie or comic book, the flat images on the screen or page seemed to come to life — in all three dimensions! You'll look great in a homemade pair of 3-D glasses, and you can make 3-D drawings to look at through them.

You will need:

a pencil

tracing paper

glue

a sheet of stiff cardboard or oaktag, at least
   nine inches square

scissors

transparent tape

two sheets of transparent plastic, one red and
   one blue (colored plastic report covers are
   perfect for this activity; you can buy them
   wherever stationery is sold)

lots of white paper

two colored pencils, one red and one light blue

With tracing paper and a pencil, trace the glasses pattern on the opposite page. Glue down the tracing paper pattern to the sheet of cardboard. Cut out the three pieces of the pattern, then cut the eyeholes out of the big piece. Tape the three pieces of the glasses together as shown. Test the fit of your glasses.

Cut a piece of red plastic two inches square. Tape it to the front of the glasses frame so that it covers an eyehole. Now tape a two-inch square of blue plastic over the other eyehole.

Trim away any extra plastic hanging over the edge of the cardboard glasses frame.

Now you have your own 3-D glasses! You can decorate them further with whatever you have on hand. Try stickers, markers, or feathers. If you've rented a 3-D video cassette, or find some 3-D comic books, you're all set. If not, you may want to make some 3-D drawings to look at.

On a sheet of white paper, draw a red line and draw a blue line right next to it. Look at them with your 3-D glasses. How does it look?

3-D drawings seem to shimmer and shift because one eye sees the red drawing and one eye sees the blue drawing. Your brain tries to put the two images together and it looks three-dimensional.

GOING FURTHER

Try drawing a 3-D cube. You can write your name in 3-D. Or draw a person with 3-D hair. Or a cat with 3-D eyes. What else can you create?

MOVING PICTURES

Have you ever walked by one of those fences that has little slits in it? The slits are so small that you can't really see what's behind the fence. If you ever walk by such a fence again, here's a way to see what's behind it: go past the fence very fast (you can run, or ride a bike or a skateboard) and look through as you go by. Suddenly, you can put together the image of what's hiding behind the fence.

Your eyes and brain have an amazing ability to hold an image for a split second after you've seen it. This ability is called *persistence of vision*. When you run by a fence, your persistence of vision makes all the tiny scenes you see through slits in the fence blend into one scene.

Movies and cartoons are made from a series of still pictures. They depend upon your persistence of vision to make them seem to move.

Before movies were invented, some people had moving picture toys. You can make some of these miraculous toys described in the next few pages to entertain yourself and your friends.